My goal in moving to Colorado was to relate my work to that landscape, to learn to paint directly from what was out there, to integrate it and make my work authentic. My painting is an amalgam of abstract painting and my study of Chinese painting, of drawings and all that I've observed in nature. I wanted to become connected to a particular landscape, so I could learn from it.

I was drawn to the pines as a motif when I was out in Colorado a couple of years ago. The actual experience of the pines was very hard to translate into drawing because they are much more complicated than even these painted versions show. It was actually looking at Chinese examples that gave me clues as to how to simplify the pines. My pines were drawn from an amalgam of what I was seeing everyday in the landscape and the Chinese stylizations. This type of pine is more prevalent in Chinese painting of the early 20th century by Wu Chang Shih and Qi Bai Shih, beginners of the modern movement. In their works the trees actually become calligraphy and are much more abstracted than the literal pines seen in older painting. That was a clue to me because it fit my more calligraphic way of painting.

While I was working the pines became an active symbol of perseverance, of standing up to the elements and patience and all of those qualities that are traditionally attributed to them. In the Chinese view of nature plants are very symbolic—bamboo is elegance, orchids are a sign of spring's arrival and prunus represents the contrasts of the seasons. Prunus branches are very gnarled, angular and dramatic, so when their flowers appear winter's austerity is banished by their beauty. By contrast, the pine tree represents an ability to maintain yourself through the austerity and difficulty of winter. They stand through all adversity, never changing, very permanent. At the end of every I-Ching reading I remember perseverance is always counseled. Hold on, like the pine, because things are always changing and what may be a dark period will turn around later, so you mustn't run away.

Brad Davis

"*Small Pine Study #1*," 1983, 32" x 38"

"*Small Pine Study #2*," 1983, 32" x 38"

When Brad Davis and his wife the painter, Janis Provisor, left New York City for their new home in Colorado they went in search of new beginnings. Recently married and filled with anticipation for the scenery they had learned to love on previous trips to the region they departed with a sense of high adventure. But the early months were not so rewarding. Settling in proved more problematic than expected, then a sequence of critical illnesses brought further complications. For Davis painting became the way through troubled times. The sturdy pines he had chosen as his motif fulfilled their ancient role as symbols of endurance. Intuition and training solidly supported him in the execution of this suite of paintings.

Coiled above a mountain peak his physique is superb,
How old is he? What Man can tell?
When the Yellow Emperor was compounding the elixir of immortality
at the beginning of time,
(The Pine's) dense scales and claws were already fully grown.

Translated inscription on the painting *Dragon Pine on Mount Huang*. *c. 1650-60*
Hung-Jen, 1610-64, Collection of the Metropolitan Museum of Art, Gift of Douglas Dillon

The challenges of a relocation were not really new. As a young painter Davis had gone to New York in the late 1960s seeking the creative stimulation and the audience that magnetically attracts Midwesterners. Trained at the Art Institute of Chicago and the University of Minnesota, Davis had mastered abstract expressionism but on arrival in New York immediately found new directions for his work. Establishing a personal style as a means of establishing an art world identity and career was the goal of his generation and Davis had both the motivation and ability. He found early appreciation for large emblematic paintings which fused intricately drawn militaristic symbols with a high-keyed palette of rainbow-hued stains. These lurid, metallic emblazoned paintings of great scale were both ironic manifestations of his disillusionment with American militarism and announcements of his entry into the combat of the artistic avant-garde. Socially and artistically political, these works also stated his interest in the power of the iconic symbols of Eastern religions. In New York, Davis was able to deepen his understanding of Oriental art through study of the city's rich museum collections.

Through these opportunities Davis acquired a particular interest in Chinese landscape painting which has been noticeable in his work since the mid-1970s, and which reaches a new plateau of excellence in *The Pines*. With a sense of purpose and commitment, Davis initiated himself as a modern apprentice to a tradition traceable to the sixth century A.D. He gradually realized that the frantic search for originality and "the New" which was going on around him was based either upon a blatant ignorance or rejection of the past. As he describes it, he could look back over his shoulder and see centuries of phenomenal artistic achievements. He found himself wanting to join the company of those artists in whom reverence for the traditions of the past is expressed through imitation of both attitude and style.

Imitation, borrowing, copying is a loaded issue in 20th century art, yet specific analysis reveals that new artists always owe some debts to their ancestors or even their contemporaries. It is a question of where the emphasis is placed and how the borrowings are used. When imitation is a propelling force leading to invigorated forms reinterpreted through new materials, the criteria of uniqueness can be waived because originality exists and, occasionally, a new movement is born. Such was the case in the mid-1970s, when a number of artists including Brad Davis began to look at traditional examples of ornament and decoration. Coming from varied artistic beginnings with differing personal motivations they shared one trait, boredom with minimal art, its rigid formality and arbitrary rejection of human sentiment. Some were feminists, some were esthetes, some were simply traditionalists, but all felt liberated when they found they had a community in which to participate. The art which they made was colorful and quirky, sometimes impermanent, sometimes monumental, but always exuberantly filled with references to organic form and vital artistic thought processes. Critics called the whole thing the "Pattern and Decoration Movement," or simply "P and D."

With sculptor Ned Smyth, Brad Davis created the landmark of "P and D" in a 1977 exhibition at the Holly Solomon Gallery. Their collaborative work entitled *The Garden*, filled the room with cast concrete colonnades and a sculptured pergola by Smyth and paintings of landscape and animals by Davis. Unique for its time as a joint venture, it was also significant for its knowledgeable blending of religious and architectural motifs from East and West presented as a natural pantheon.

Throughout this time Davis was intensifying his command of the conventional language of Oriental art. Some commercial success enabled him to begin collecting Chinese paintings, and practice with brush and pen increased his facility at rendering complex floral and landscape motifs. It should be stated here that Davis has never formally studied Chinese painting or calligraphy as taught by masters residing in Manhattan. He says with emphasis that he is not trying to be a Chinese painter. What he seeks is the confidence and assurance to depict organic forms spontaneously. Owning up to his formative years as a student of abstract expressionism with Peter Busa, second-generation member of the New York School, Davis subconsciously reveals a fundamental interest in action painting. Yet his love of landscape and nature is so deeply rooted that he cannot submerge that feeling in pure abstraction and needs some form of representational depiction to express what he feels. The Chinese painters are ideal models offering both reverence for nature and the freedom from faithful representation which has characterized the dominant schools of Western landscape painting.

Davis acknowledges the gifts of the Hudson River School painters and of Frederick Church and Albert Bierstadt, who visited the West, confronted nature and exalted its spaciousness within their canvases. Philosophically he mourns the loss of innocence, the rise of cynicism that prevents contemporary painters from taking a similarly awestruck stance in the presence of magnificent scenery. Unable to accommodate the 19th century viewpoint, feeling too urbanized, even politicized, Davis, a contemporary painter with an interest in landscape, fuses the historic Chinese conception of abstract foliage with an abstract facture developed in the New York School, c. 1955.

The invitation for an exhibition at the Taft Museum had been proposed in the spring of 1983, before Davis moved to Colorado. Stopping in Cincinnati on the drive west, Davis surveyed the gallery, determining how many paintings would be needed. Once settled in a new studio in Carbondale, Davis ordered canvas stretchers in multiples of preferred sizes. Davis likes to work on sets of paintings, pairs, groups of three, four, even five. With a given theme or motif in mind too many ideas would crowd into one canvas, so a second, third or fourth is always developing simultaneously, or maybe a bit behind. Further on, larger, more ambitious canvases result from the experience of smaller works or from drawings. Talking about his working method, Davis projects an image of efficiency, even economy. Not that he's mechanical or frugal, he simply isn't wasteful of ideas or materials. And he is building a body of work with consistency and concentration.

"*2 Pines Study #1*," 1984, 52⅛" x 56"

"*2 Pines Study #2*," 1984, 52⅛" x 56"

Selecting the subject of the pines, Davis started the small pair and the large single canvas, "*3 Pines in the Snow*." As winter advanced he noticed how the craggy foothills and distant mountains were softening into gently rolling mounds under the blanket of snow. Against the blue-white terrain, Davis' three pines gyrate in a chilly choreography. Animated by flashing brushstrokes and brilliant color, this painting is visibly more controlled, more carefully organized, projecting conscious authority and a cooler mood than the *Small Pine Studies*. In these Davis was exploring, taking risks with a bolder palette, fingerpainting, splattering and dashing on paint while concentrating on a single trunks with sparse branches. A new brand of acrylic paint Davis has recently found enables him to use paint loosely, wet into wet, without diminishing the color saturation of separate hues.

Clearly the gestures of the Colorado pinions have called forth a more gestural style of painting that is new in Davis' art. Previously his paintings were animated by a variety of floral motifs and a wealth of carefully structured and detailed naturalistic observations. Paintings involving animals were mildly anecdotal, reminiscent of forgotten fables. Sometimes flamboyant when seeking to be tropical or perverse when using polyester borders as a pluralistic pun on Chinese silks, Davis narrowly missed tripping over the tight wire of good taste. Stretching that wire was a basic issue to the "P and D" artists and Davis did not avoid it. But now he has put it all behind him, making paint and painting the central principles of his art.

The pair titled "*2 Pines Studies*" make the shift even more apparent. Moving out from the smaller studies, in this pair Davis chose to work with stronger contrasts in the theme. Painting the variants simultaneously, striving to evoke different feelings, they became distinct in his mind. And in the "*2 Pines Studies*" Davis also began to work with variations of surface texture. Acrylic paint cannot produce *impasto*, but using his new acrylics Davis found he could build a denser surface. Seeing them initially one thinks of deKooning; Davis agrees, adding Jackson Pollock's name too. He is right, of course, because the marbling, skeining, paint technique was pioneered in Pollock's poured on canvases. Against these fields, the pine branches of "*2 Pines Study # 2*" spring forward in a torqued gesture that is rhythmic and violent. Smaller clusters of needles splay outwards as if blown in an atmospheric maelstrom that is passionate and tumultuous. "*2 Pines Study # 1*" is less agitated, though by no means serene. The pinions now stand in tense, opposing combat.

Mountain Falls and *Pine Shore* are a vertically oriented pair that, while standing within the context of the series, introduce an expanded landscape repertoire. Rushing waters pouring over rocks dominate the canvases, and in *Mountain Falls* the pinion motif is even eliminated. Davis began this painting early in the course of his activity. Planning the series he had known that the largest canvas, *Pine by the River*, would be the climax of the suite and he wanted to bring the theme of a moving stream into the working process. In *Mountain Falls* a cascade spills through rocky ledges forming whitewater sprays and dark pools. The landscape forms, immediately apprehensible for their directness and vigor, rivet our attention by the constant dialectic established between representational and abstract qualities. Yes, it is paint, but it is also foaming water; yes, those are rocks, but they are also boxy, drawn shapes on a flat surface. Here color has significantly heightened the tension of the dualisms. Making reference to the blue and gold style of Northern Sun painting, Davis has limited his palette and made key colors dominate throughout the composition. No atmospheric blendings blur the insistent vertical structure that subtley insists on the verifiable flatness of the paintings' support. *Mountain Falls* has a frontal, nearly symmetrical composition; *Pine Shore* offers its contrast for it is more assymmetrical with stronger allusions to a deeper space. Waters here must course around an upright, rock outcropping that sprouts a pine whose branches are thrown into the middle ground. Stubby rocks in descending ranks break waters highlighted by broad brisk brushmarks. *Pine Shore* has the same eccentrically limited palette enlivened by an especially remarkable tangerine hue.

It is evident that Davis has a unique perception of the color of the Colorado landscape. These are studio paintings, remembered evocations of what he has seen fused with what he has learned about the emotional power of color. Working in a temporary studio, without windows and hence without natural illumination, the color he has used is intuited from vivid, deeply felt recollections of sensations gathered on wanderings in the countryside. *Pine by the River* is illuminated by an eerie light used to enhance the feeling of a rushing turbulence that draws the viewer into a vertiginous relationship with the fast paced movement of a mountain stream. No resting place is offered as the eye sweeps back and fourth along the horizontal surface. Indications of a rocky ledge at midpoint which might serve as a way station are countered by an upward tilt that offers no foothold and by the pine that throws itself forward into the viewer's space. Once again, emphasis has been placed upon colorful, calligraphic bravura effects that overwhelm naturalistic speculations while confirming artistic mastery.

Midnight Pine is the coda to the series. A single, sinuous branch diagonally displayed against a murky violet stillness. Moonlight picks out the gnarled limb and illuminates its needles; the atmosphere glows with flickering motes of paint. Pensive, even melancholy, *Midnight Pine* brings the sequence to a state of rest, not a closure. Much has been stated, more remains possible.

Ruth K. Meyer, Director
The Taft Museum

Pine trees are like people of high principles whose manner reveals an inner power. They resemble young dragons coiled in deep gorges; they have an attractive and graceful air, yet one trembles to approach them, awed by the hidden power ready to spring forth. Those who paint pines should keep this meaning in mind. The brush will then effortlessly produce extraordinary results.

Pines painted by Ma Yuan were lean and strong as iron.

From the "Book of Trees," *Chieh Tzu Yuan Hua Chuan, (The Mustard Seed Garden Manual of Painting)*, 1679.

"Pine by the River," 1984, 47½" x 120"

Shore," 1984, 72'' x 48''

"Mountain Falls," 1983, 72'' x 48''

BRAD DAVIS

Born in Duluth, Minnesota, 1942
Studied at the University of Chicago (1961-62);
School of The Art Institute of Chicago (1962-63);
University of Minnesota, Minneapolis (B.A., 1966);
Hunter College, New York (1968-70)
Lives in New York and Carbondale, Colorado

Selected One-Artist Exhibitions

1972 98 Greene Street Loft, New York
1975 Holly Solomon Gallery, New York
1977 Holly Solomon Gallery, New York (collaboration with Ned Smyth)
1979 Holly Solomon Gallery, New York
1980 Toni Birckhead Gallery, Cincinnati
The Mayor Gallery, London
1981 Holly Solomon Gallery, New York
1982 Holly Solomon Gallery, New York (collaboration with Ned Smyth)
1983 Holly Solomon Gallery, New York

Selected Group Exhibitions

1966 University Gallery, University of Minnesota, Minneapolis, "Some Young Minnesota Artists"
Walker Art Center, Minneapolis, "Two Person Show"
1972 Whitney Museum of American Art, New York, "1972 Annual Exhibition: Contemporary American Painting"
1973 Whitney Museum of American Art, New York, "American Drawings 1963-1973"
1975 The Clocktower, Institute for Art and Urban Resources, New York, "Artists' Toys"
Art Museum of South Texas, Corpus Christi, "Purchase Show"
1977 The American Foundation for the Arts, Miami, Miami, "Patterning and Decoration" (traveled to Rice University, Houston)
P.S. 1, Institute for Art and Urban Resources, Long Island City, New York, "A Painting Show"
Sarah Lawrence College Gallery, Bronxville, New York, "Painting 75 76 77" (traveled to Miami and Cincinnati). Catalogue.
1980 Mannheimer Kunstverein, Mannheim, West Germany, "Dekor" (traveled to Amerika House, West Berlin; Museum of Modern Art, Oxford, England)
Neue Galerie/Sammlung Ludwig, Aachen, West Germany, "Les Nouveaux Fauves—Die Neuen Wilden." Catalogue.
United States Pavilion, 39th Venice Biennale, Italy, "Drawings: The Pluralist Decade" (traveled to the Institute of Contemporary Art, University of Pennsylvania, Philadelphia). Catalogue.
1981 The New Museum, New York, "Alternatives in Retrospect: An Historical Overview 1969-1975"
Sidney Janis Gallery, New York, "New Directions: Contemporary American Art from the Commodities Corporation Art Collection"
Squibb Gallery, E. R. Squibb and Sons, Inc., Princeton, New Jersey, "Aspects of Post-Modernism: Decorative and Narrative Art"
1982 Kestner-Gesellschaft, Hannover, West Germany, "New York Now" (traveled to Munich; Lausanne, and Düsseldorf). Catalogue.
The Hudson River Museum, Yonkers, New York, "Ornamentalism: The New Decorativeness in Architecture and Design"
1983 The Berkshire Museum, Pittsfield, Massachusetts, "New Decorative Art"
Kunstmuseum, Lucerne, Switzerland, "Back to the U.S.A.: Amerikanische Kunst der Siebziger und Achtziger" (traveled to Bonn, and Stuttgart)
The Museum of Modern Art, New York, "Some Recent Acquisitions"
1984 Whitney Museum of American Art "Five Painters in New York." Catalogue.

Selected Bibliography

Anderson, Alexandra. "Clay Gardens." *Portfolio*, 4 (March-April 1982), pp. 17-18.

Davis, Douglas, "Portrait of a Young Artist." *Time*, 99 (February 7, 1972), p. 79.

deAk, Edit. In "Reviews," *Artforum*, 16 (December 1977), p. 63.

French-Frazier, Nina. In "New York Exhibitions." *Art International*, 23 (December 1979), pp. 39-40.

Goldin, Amy. *Patterning and Decoration* (exhibition catalogue). Miami, Florida: The American Foundation for the Arts, 1977.

Grove, Nancy. "Brad Davis." *Arts Magazine*, 50 (December 1975), p. 13.

Honnef, Klaus. "New York Aktuelle." *Kunstforum International*, 61 (May 1983). pp. 37, 112-16.

Hunter, Sam. "Post-Modernist Painting." *Portfolio*, 4 (January-February 1982), pp. 46-53.

Jensen, Robert, and Patricia Conway. *Ornamentalism: The New Decorativeness in Architecture and Design*. New York: Clarkson N. Potter, 1982, pp. 20, 241, 251-53.

Larson, Kay. "Art," *The Village Voice*, September 24, 1979, p. 87.

Lawson, Thomas. "Brad Davis." *Flash Art*, no. 92-93 (October-November 1979), p. 23.

Lucie-Smith, Edward. *Art in the Seventies*, Ithaca, New York: Phaidon Books/Cornell University Press, 1980.

Morgan, Stuart. "Brad Davis at the Mayor Gallery, London." *Artscribe* (December 1980), pp. 42-43.

Oliva, Achille Bonito. "The Bewildered Image." *Flash Art*, no. 96-97 (March-April 1980), pp. 32-35.

________. "Transavantgarde." *Interarte 20*, May 1981, pp. 5-7, 16.

Rickey, Carrie. "Decoration, Ornament, Pattern and Utility: Four Tendencies in Search of a Movement." *Flash Art*, no. 90-91 (June-July 1979), pp. 19-23.

Tatransky, Valentin. "Brad Davis/Ned Smyth." *Arts Magazine*, 52 (November 1977), p. 28.

Tomkins, Calvin. "Matisse's Armchair." *The New Yorker*, 56 (February 25, 1980), pp. 108-13.

Yau, John. "Brad Davis and Ned Smyth at Holly Solomon." *Art in America*, 66 (January-February 1978), 114-15.